Palette of Color Monograph Series

The Chemistry of Vat Dyes

by Dianne N. Epp

Series Editor
Mickey Sarquis, Director
Center for Chemical Education

Terrific Science Press
Miami University Middletown
Middletown, Ohio

Contents

◼ Acknowledgments

The author and editor wish to thank the following individuals who have contributed to the development of *The Chemistry of Vat Dyes.*

Terrific Science Press Design and Production Team
Susan Gertz, Amy Stander, Lisa Taylor, Thomas Nackid, Stephen Gentle

Terrific Science Press Laboratory Testing Coordinator
Andrea Nolan

Reviewers

David H. Abrahams	Dexter Chemical Corporation	New York, NY
Susan Hershberger	Miami University	Oxford, OH

◼ Center for Chemical Education Staff

Mickey Sarquis, Director
Bruce L. Peters, Jr., Associate Director
Billie Gerzema, Administrative Assistant

Assistants to Director

Susan Gertz	Mark Sabo
Lynn Hogue	Lisa Meeder Turnbull

Project Coordinators and Managers

Richard French	Andrea Nolan
Betty Kibbey	Ginger Smith
Carl Morgan	

Research Associates and Assistants

Kersti Cox	Bronwyn Nelson
Stephen Gentle	Michael Parks
Susan Hershberger	Amy Stander
Robert Hunter	Lisa Taylor
Anne Munson	David Tomlin
Thomas Nackid	

Program Secretaries

Victoria Burton	Ruth Willis
Debra Chan	

Graduate Assistants

Michelle Diebolt	Richard Rischling
Nancy Grim	Michella Stultz

Dedication

To my best friend and husband, Tony, who moved to Ohio with me and without whose encouragement and support this could not have happened

and

To my chemical education models, mentors, and friends, Mickey and Jerry Sarquis, who gave me the opportunity to try my hand at a different kind of "teaching."

Foreword

The Chemistry of Vat Dyes is the first in a three-volume series of monographs entitled *Palette of Color*. This series is aimed at enabling high school chemistry teachers to introduce their students to a fascinating area of industrial chemistry—dyes and colorants. These monographs provide background information on the history and chemistry of various dyes and colorants, together with hands-on activities on producing, testing, and using these chemicals. Dianne Epp has brought together an excellent collection of chemistry activities in a format that is convenient and easy to use. As in all volumes published by Terrific Science Press, each of these activities has been tested by teachers in the Center for Chemical Education's (CCE) Terrific Science Programs and reviewed by experts in the field to ensure accuracy, safety, and pedagogical effectiveness. We believe that these monographs will enhance the relevance and appeal of the high school chemistry teacher's repertoire.

Dianne created these volumes while on a 1994–95 sabbatical from East High School in Lincoln, NE. During that year, Dianne joined the CCE team as a Teacher Fellow and worked on other curriculum-development efforts, including integrating microscale laboratory activities into Miami University's general chemistry curriculum. We want to thank Dianne for sharing her keen insights into the topics of dye chemistry, microscale chemistry, and chemical education; for her hard work in developing this series; and for allowing us to publish it.

In addition to the *Palette of Color* monograph series, the CCE offers many other science education opportunities and resource materials for teachers and students at all levels. A summary of these opportunities is outlined on Pages viii–xi. While each initiative within the Center has a unique focus and addresses the needs of a distinct population, all programs emphasize hands-on, inquiry-based chemical education through which students develop their abilities to work together to solve scientific challenges, think critically, and utilize their powers of observation.

We hope you will find that these monographs provide you with a useful and exciting way to involve your students in doing chemistry through integrated real-world themes. We welcome your comments at any time and are interested in learning about especially successful uses of these materials.

Mickey Sarquis, Director
Center for Chemical Education
March 1995

◼ The Center for Chemical Education

Built on a tradition of quality programming, materials development, and networking between academia and industry, Miami University's Center for Chemical Education (CCE) encompasses a multifaceted collaboration of cross-grade-level and interdisciplinary initiatives begun in the mid-1980s as Terrific Science Programs. These initiatives are linked through the centrality of chemistry to the goal of fostering quality hands-on, minds-on science education for all students. CCE activities include credit coursework and other opportunities for educators at all levels; K–12 student programs; undergraduate, graduate, and postgraduate programs in chemical education; materials development, including teacher resource materials, program handbooks, and videos; public outreach efforts and networking to foster new and existing partnerships among classroom teachers, university-based science educators, industrial scientists, and professional societies.

Professional Development for Educators

Credit Courses

The Center for Chemical Education offers a variety of summer and academic-year workshop-style courses for K–12 and college teachers. While each workshop has a unique focus, all reflect current pedagogical approaches in science education, cutting-edge academic and industrial research topics, and classroom applications for teachers and students. Short courses provide opportunities for educators to enrich their science teaching in a limited amount of time. All courses offer graduate credit.

Non-Credit Courses

Academies allow CCE graduates and other teachers to attend special one-day sessions presented by leading science educators from around the United States. Offerings include seminars, mini-workshops, and share-and-swap sessions.

Internships

Through 8- to 10-week summer internships, program graduates work as members of industrial teams to gain insight into the day-to-day workings of industrial laboratories, enabling them to bring real-world perspectives into the classroom.

Fellowships

Master teachers at primary, secondary, and college levels do research in chemical education and undertake curriculum and materials development as Teacher Fellows with the Center for Chemical Education. Fellowships are available for the summer and the academic year.

K–12 Student Programming

Summer Camps

A variety of summer camps are available to area elementary, middle, and high school students. These camps not only provide laboratory-based enrichment for students, but also enable educators in summer courses to apply their knowledge of hands-on exploration and leadership skills. Satellite camps are offered at affiliated sites throughout the country.

Science Carnivals

Carnivals challenge elementary school students with hands-on science in a non-traditional atmosphere, encouraging them to apply the scientific method to activities that demonstrate scientific principles. Sponsoring teachers and their students host these carnivals for other students in their districts.

Palette of Color—The Chemistry of Vat Dyes

Super Saturday Science Sessions	High school students are introduced to industrial and research applications of science and technology through special Saturday sessions that involve the students in experiment-based problem-solving. Topics have included waste management, environmental sampling, engineering technology, paper science, chemical analysis, microbiology, and many others.
Ambassador Program	Professional chemists, technicians, and engineers, practicing and recently retired, play important roles as classroom ambassadors for high school and two-year college students. Ambassadors not only serve as classroom resources, but they are also available as consultants when a laboratory scenario calls for outside expertise; they mentor special projects both in and out of the classroom; and they are available for career counseling and professional advice.

Undergraduate and Graduate Student Programming

Teaching Science with TOYS Undergraduate Course	This undergraduate course replicates the Teaching Science with TOYS teacher inservice program for the preservice audience. Students participate in hands-on physics and chemistry sessions.
General Chemistry Initiative	This effort is aimed at more effectively including chemical analysis and problem solving in the two-year college curriculum. To accomplish this goal, we are developing and testing discovery-based laboratory scenarios and take-home lecture supplements that illustrate topics in chemistry through activities beyond the classroom. In addition to demonstrating general chemistry concepts, these activities also involve students in critical-thinking and group problem-solving skills used by professional chemists in industry and academia.
Chemical Technology Curriculum Development	Curriculum and materials development efforts highlight the collaboration between college and high school faculty and industrial partners. These efforts will lead to the dissemination of a series of activity-based monographs, including detailed instructions for discovery-based investigations that challenge students to apply principles of chemical technology, chemical analysis, and Good Laboratory Practices in solving problems that confront practicing chemical technicians in the workplace.
Other Undergraduate Activities	The CCE has offered short courses/seminars for undergraduates that are similar in focus and pedagogy to CCE teacher/faculty enhancement programming. In addition, CCE staff members provide Miami University students with opportunities to interact in area schools through public outreach efforts and to undertake independent study projects in chemical education.
Degree Program	Miami's Department of Chemistry offers both a Ph.D. and M.S. in Chemical Education for graduate students who are interested in becoming teachers of chemistry in situations where a comprehensive knowledge of advanced chemical concepts is required and where acceptable scholarly activity can include the pursuit of Chemical Education research.

Educational Materials

The Terrific Science Press publications have emerged from CCE's work with classroom teachers of grades K–12 and college in graduate-credit, workshop-style inservice courses. Before being released, our materials undergo extensive classroom testing by teachers working with students at the targeted grade level, peer review by experts in the field for accuracy and safety, and editing by a staff of technical writers for clear, accurate, and consistent materials lists and procedures. The following is a list of Terrific Science Press publications to date.

Science Activities for Elementary Classrooms (1986)

Science SHARE is a resource for busy K–6 teachers to enable them to use hands-on science activities in their classrooms. The activities included use common, everyday materials and complement or supplement any existing science curriculum. This book was published in collaboration with Flinn Scientific, Inc.

Fun With Chemistry Volume 2 (1993)

The second volume of a set of two hands-on activity collections, this book contains classroom-tested science activities that enhance teaching, are fun to do, and help make science relevant to young students. This book was published in collaboration with the Institute for Chemical Education (ICE), University of Wisconsin-Madison.

Polymers All Around You! (1992)

This monograph focuses on the uses of polymer chemistry in the classroom. It includes several multi-part activities dealing with topics such as polymer recycling and polymers and polarized light. This monograph was published in collaboration with POLYED, a joint education committee of two divisions of the American Chemical Society: the Division of Polymer Chemistry and the Division of the Polymeric Materials: Science and Engineering.

Santa's Scientific Christmas (1993)

In this school play for elementary students, Santa's elves teach him the science behind his toys. The book and accompanying video provide step-by-step instructions for presenting the play. The book also contains eight fun, hands-on science activities to do in the classroom.

Teaching Chemistry with TOYS Teaching Physics with TOYS (1995)

Two volumes contain more than 75 chemistry and physics activities for grades K–9 and were developed in collaboration with and tested by classroom teachers from around the country. These volumes were published in collaboration with McGraw-Hill, Inc.

Palette of Color Monograph Series (1995)

The three monographs in this series presents the chemistry behind dye colors and shows how this chemistry is applied in "real-world" settings:
- The Chemistry of Vat Dyes (1995)
- The Chemistry of Natural Dyes (scheduled)
- The Chemistry of Food Dyes (scheduled)

Science in Our World Teacher Resource Modules (scheduled 1995)

Each volume of this six-volume set presents chemistry activities based on a specific industry—everything from pharmaceuticals to metallurgy. Developed as a result of the *Partners for Terrific Science* program, this set explores the following topics and industries:
- Chemistry and the Food Industry (Procter & Gamble)
- Chemistry of Metals (Armco Inc.)

- Institutional Products Chemistry (Diversey Corporation)
- Oleochemistry (Henkel Corporation—Emery Group)
- Pharmaceutical Chemistry (Marion Merrell Dow Inc.)
- Polymer Chemistry (Quantum Chemical Company)

Teaching Science with TOYS Teacher Resource Modules (scheduled 1995) The modules in this series are designed as instructional units focusing on a given theme or content area in chemistry or physics. Built around a collection of grade-level-appropriate TOYS activities, each Teacher Resources Module also includes a content review and pedagogical strategies section. Topics to be released in 1995 include the following:
- Mechanical Energy and Energy Conversions (intermediate)
- Linear Motion (primary)
- Using Your Senses to Explore Matter (elementary)
- States of Matter and Changes of State (middle school)

Terrific Science Network

Affiliates College and district affiliates to CCE programs disseminate ideas and programming throughout the United States. Program affiliates offer support for local teachers, including workshops, resource/symposium sessions, and inservices; science camps; and college courses.

Outreach On the average, graduates of CCE professional development programs report reaching about 40 other teachers through district inservices and other outreach efforts they undertake. Additionally, graduates, especially those in facilitator programs, institute their own local student programs. CCE staff also undertake significant outreach through collaboration with local schools, service organizations, professional societies, and museums.

Newsletters CCE newsletters provide a vehicle for network communication between program graduates, members of industry, and other individuals active in chemical and science education. Newsletters contain program information, hands-on science activities, teacher resources, and ideas about how to integrate hands-on science into the curriculum.

For more information about any of the CCE initiatives, contact us at

Center for Chemical Education
4200 East University Blvd.
Middletown, OH 45042
(513) 424-4444 x378
FAX (513)424-4632
e-mail: CCE@miavx3.muohio.edu.

What Are the *Palette of Color* Monographs?

Look around you—color is everywhere. The clothes we wear, the food we eat, the posters with which we decorate our rooms; indeed all of our surroundings, natural and man-made, abound with color. From prehistoric times people have been fascinated with color; from cave paintings to the latest computers, color has been our constant companion.

The *Palette of Color* monograph series enables high school chemistry teachers to challenge students to explore the chemistry behind dyes. Each monograph examines a different class of dyes and investigates the chemistry using principles common to most high school chemistry curricula. Problem-solving and inquiry-based activities involve students in answering questions posed about the dyes and their uses.

The Chemistry of Vat Dyes

Indigo and inkodyes are used to illustrate how vat dyes are synthesized and used. Until the end of the 19th century, all colors were obtained from natural sources, but today the number of synthetic colorants exceeds 7,000. One class of these colorants, the vat dyes, contains not only the oldest natural dyes known, but also many important synthetic dyes. This class of dyes is studied in this monograph.

The Chemistry of Natural Dyes

For thousands of years, dyes were obtained from natural sources, such as plants and animals. In spite of the fact that synthetic dyes have replaced many natural dyes for commercial use, natural dyes still hold a fascination and are used extensively by artisans around the world. This monograph investigates the most common type of natural dyes, known as acid or anionic dyes, and their reaction with wool.

The Chemistry of Food Dyes

Dyes aren't just for fabrics—colorants have been added to food for centuries to enhance its appearance. This monograph investigates both the compounds which give foods their natural color and the synthetic colorants currently approved for use in foods.

How to Use *The Chemistry of Vat Dyes*

The *Palette of Color* monographs are intended for use by a secondary chemistry teacher. A monograph may be inserted into the curriculum when the appropriate chemistry concept is being examined as a practical application of that concept. Monographs might also be used as independent study units for students.

Each monograph is organized in two parts: Teacher Background Information and Classroom Materials. The Teacher Background Information section includes a review of pertinent science content, notes and setups for the activities, and cross-curricular activities to supplement the science activities. The Classroom Materials section includes student background and activity handouts and overheads.

What's in *The Chemistry of Vat Dyes*

Until the end of the 19th century, all colors were obtained from natural sources, but today the number of synthetic colorants exceeds 7,000. One class of these colorants, the vat dyes, contains not only the oldest natural dyes known, but also many important synthetic dyes. This class of dyes is studied in this monograph.

What Students Do

In this monograph, students do the following:

- study the oxidation-reduction chemistry of vat dyes;
- investigate the construction of denim as an example of a vat-dyed fabric;
- prepare a sample of synthetic indigo;
- prepare an indigo-dye bath to experiment with the vat-dye process;
- study the characteristics of indigo as a vat dye;
- determine experimentally the optimum conditions for developing the color of Inkodye, a leuco-base vat dye; and
- compare the abilities of natural and synthetic fibers to absorb Inkodye.

Key Ideas

The following key ideas are covered:

- vat dyes as examples of oxidation/reduction chemistry;
- solubility as a function of chemical composition;
- structures of natural polymers, such as cellulose, compared to synthetic polymers, such as polyesters; and
- commercial processes used to dye fabrics.

Time Frame

Teachers should select from the variety of activities to fit the laboratory time available. A minimum of two 50-minute lessons would be needed to cover the topic of vat dyes.

Employing Appropriate Safety Procedures

Experiments, demonstrations, and hands-on activities add relevance, fun, and excitement to science education at any level. However, even the simplest activity can become dangerous when the proper safety precautions are ignored or when the activity is done incorrectly or performed by students without proper supervision. While the activities in this book include cautions, warnings, and safety reminders from sources believed to be reliable and while the text has been extensively reviewed by classroom teachers and university scientists, it is your responsibility to develop and follow procedures for the safe execution of the activities you choose to do and for the safe handling, use, and disposal of chemicals in accordance with local and state regulations and requirements.

Safety First

- Collect and read the Materials Safety Data Sheets (MSDS) for all of the chemicals used in your experiments. MSDS's provide physical property data, toxicity information, and handling and disposal specifications for chemicals. They can be obtained upon request from manufacturers and distributors of these chemicals. In fact, MSDS's are often shipped with the chemicals when they are ordered. These should be collected and made available to students, faculty, or parents for information about specific chemicals used in these activities.

- Read and follow the American Chemical Society Minimum Safety Guidelines for Chemical Demonstrations on the next page. Remember that you are a role model for your students—your attention to safety will help them develop good safety habits while assuring that everyone has fun with these activities.

- Read each activity carefully and observe all safety precautions and disposal procedures. Determine and follow all local and state regulations and requirements.

- Never attempt an activity if you are unfamiliar or uncomfortable with the procedures or materials involved. Consult a college or industrial chemist for advice or ask him or her to perform the activity for your class. These people are often delighted to help.

- Always practice activities yourself before using them with your class. This is the only way to become thoroughly familiar with an activity, and familiarity will help prevent potentially hazardous (or merely embarrassing) mishaps. In addition, you may find variations that will make the activity more meaningful to your students.

- You, your assistants, and any students participating in the preparation for or doing of the activity must wear safety goggles if indicated in the activity and at any other time you deem necessary.

- Special safety instructions are not given for everyday classroom materials being used in a typical manner. Use common sense when working with hot, sharp, or breakable objects. Keep tables or desks covered to avoid stains. Keep spills cleaned up to avoid falls.

ACS Minimum Safety Guidelines for Chemical Demonstrations

This section outlines safety procedures that must be followed at all times.

Chemical Demonstrators Must:

1. know the properties of the chemicals and the chemical reactions involved in demonstrations presented.

2. comply with all local rules and regulations.

3. wear appropriate eye protection for all chemical demonstrations.

4. warn the members of the audience to cover their ears whenever a loud noise is anticipated.

5. plan the demonstration so that harmful quantities of noxious gases (e.g., NO_2, SO_2, H_2S) do not enter the local air supply.

6. provide safety shield protection wherever there is the slightest possibility that a container, its fragments, or its contents could be propelled with sufficient force to cause personal injury.

7. arrange to have a fire extinguisher at hand whenever the slightest possibility for fire exists.

8. not taste or encourage spectators to taste any non-food substance.

9. not use demonstrations in which parts of the human body are placed in danger (such as placing dry ice in the mouth or dipping hands into liquid nitrogen.

10. not use "open" containers of volatile, toxic substances (e.g., benzene, CCl_4, CS_2, formaldehyde) without adequate ventilation as provided by fume hoods.

11. provide written procedure, hazard, and disposal information for each demonstration whenever the audience is encouraged to repeat the demonstration.

12. arrange for appropriate waste containers for and subsequent disposal of materials harmful to the environment.

Revised 6/4/88. Copyright © 1988, ACS Division of Chemical Education, Inc. Permission is hereby granted to reprint or copy these guidelines providing that no changes are made and they are reproduced in their entirety.

Part A: Teacher Background Information

What Are Vat Dyes?

Notes and Setups for Activities

Supplementary Activities

References

What Are Vat Dyes?

What Is a Dye?

Colorants for textiles include dyes and pigments. Dyes are organic chemicals which selectively absorb and reflect wavelengths of light within the visible spectrum. Dyes usually diffuse into the interior of a fiber from a water solution. Pigments are water-insoluble, microscopic-sized color particles that are usually held to the surface of a fiber by a resin. In this monograph we limit our investigation to the use of dyes as colorants.

Dye molecules vary greatly in composition and behavior. A typical dye molecule contains at least three unique chemical groups, each responsible for a particular property of the dye. The chromophore is the color-producing portion of the dye molecule, the auxochrome influences the intensity of the dye and provides a site at which the dye can chemically bond to the fabric, and the solubilizing group allows the dye molecule to be water soluble so that it is capable of interacting with a fiber in a water bath. These chemical groups include the typical examples shown in Figure A1.

Chromophores	Auxochromes	Solubilizing Groups
	$-CH_3$	$SO_3^- \ Na^+$
	$-O-CH_3$	$NH_2^+ \ Cl^-$
	$-OH$	$SO_2^- \ NH_2^+$
	$-NH_2$	$O^- \ Na^+$
$-N{=}N-$	$-NO_2$	
$-N{=}N-$ with O	$-SO_3Na$	
$C{=}C$		

Figure A1: Typical chemical groups of dye molecules

Chemical Reactions of Vat Dyes

Dyes may be classified in a number of ways, including the chemical constitution of the dye and its method of application to the fabric. Indigo, possibly the oldest dyestuff known, belongs to a class of dyes known as vat dyes. The term "vat" refers to the vessel originally used to ferment the indigo leaves. Fermentation is a process used to remove oxygen from the compound. (This is a reduction process).

Vat dyes are water-insoluble dyes which can be reduced in the presence of a base to form a water-soluble dye known as a leuco base. Vat dyes embrace a wide variety of structural types and often the molecules are quite complex. However, all vat dyes contain one or more carbonyl groups (C=O). The carbonyl group is reduced to the sodium salt by treatment with a reducing agent in the presence of an alkali. This process, discussed later in this section and shown in Figure A4, produces the water-soluble leuco-base form of the dye which can be applied to fiber using a water bath.

Cellulose fibers, such as cotton, show a significant affinity for these leuco-base dyes. This affinity arises from two factors: 1) the dye molecules have the correct geometry to fit into the relatively large pores of the cellulose fiber, and 2) the hydrogen bonds and Van der Waals forces attract the leuco base to the cellulose.

Upon exposure to oxygen (in air) in the presence of heat or light, the dye molecule oxidizes and is converted back to its insoluble form. By this time, however, the dye molecule is trapped inside the cellulose fiber and is not easily removed because of its immense size. (See Figure A2.) This trapping and insolubilizing of the large dye molecule within the cellulose structure gives the fabrics dyed using this method unusually high resistance to fading. Vat dyes, for example, are much more colorfast than dyes which attach only by reacting with the surface of a fiber. It is interesting to note that the color of the reduced dye (or leuco base) often differs from the color produced when the dye is oxidized. Indigo, for example, is yellow in the dye pot and does not acquire its characteristic deep blue until it is oxidized when exposed to air.

vat dye (insoluble) $\xrightarrow{\text{reducing agent}}$ leuco-base form (soluble) $\xrightarrow{\text{oxidation by air and heat or light}}$ vat dye trapped in fiber (insoluble)

Figure A2: General process of solubilizing a vat dye

Inkodye is a vat dye that has already been converted into a leuco-base form. This allows it to be painted directly on fabric and the colors developed by heat. Exposure to direct sunlight offers adequate heat for this process. This can even be carried out, although somewhat more slowly, by sunlight through a window. Alternatively, development may be done by ironing the damp fabric or baking the fabric in a 275°F oven.

Synthetic vat dyes generally are of two types, those derived from indigo and those derived from anthraquinone. (See Figure A3.)

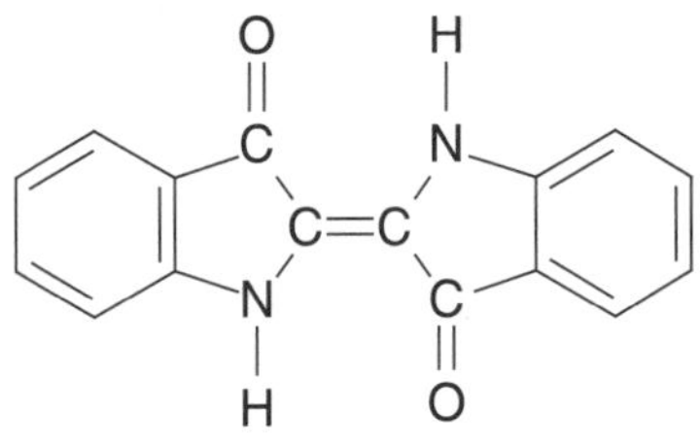

indigo 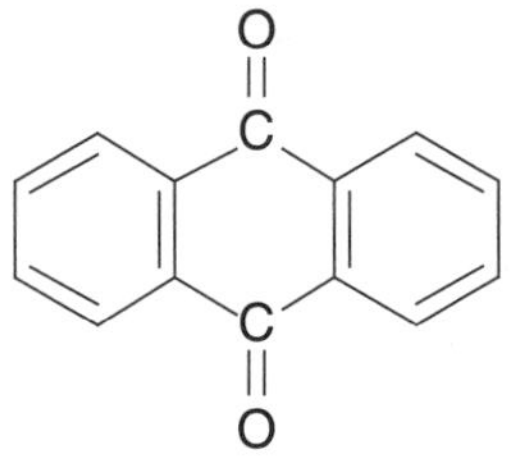   anthraquinone

Figure A3: The structures of indigo and anthraquinone

An example of the reduction of each of these chromophore types is given in Figure A4.

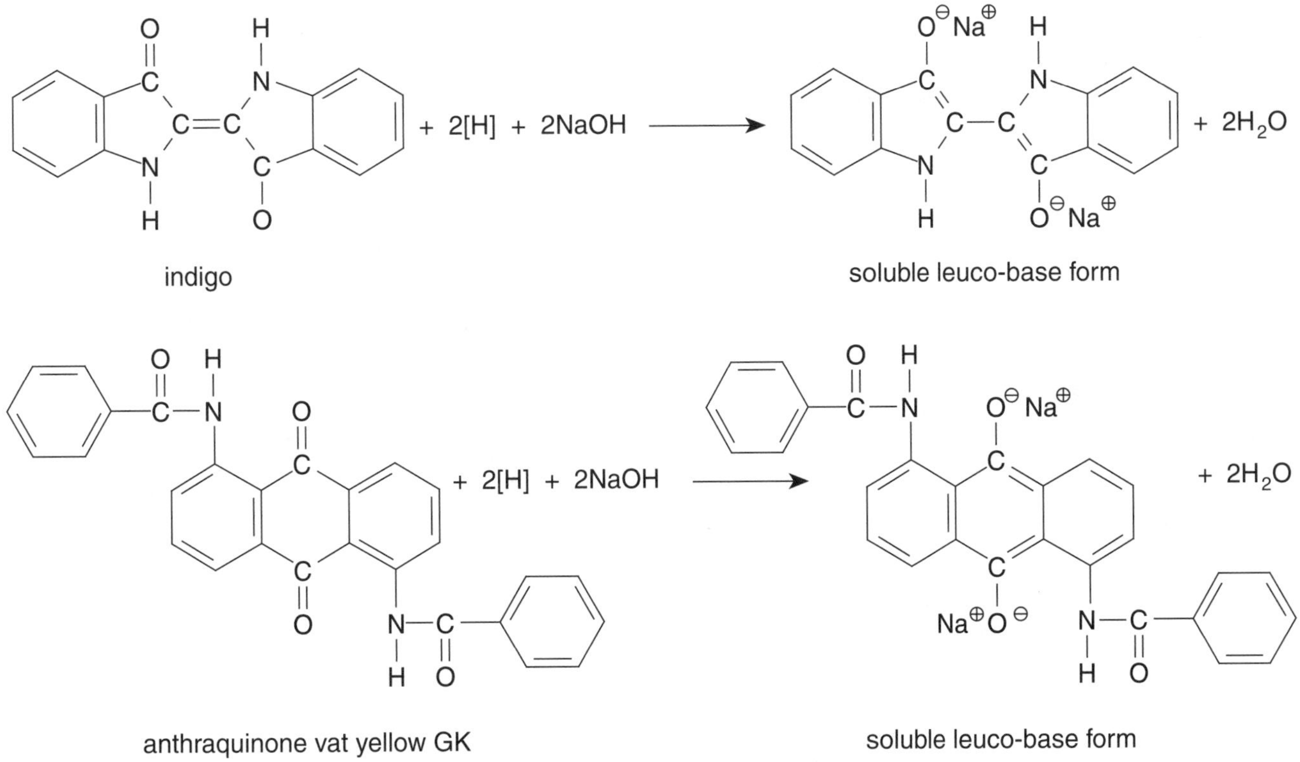

indigo

soluble leuco-base form

+ 2[H] + 2NaOH ⟶ + 2H$_2$O

anthraquinone vat yellow GK

soluble leuco-base form

Figure A4: Reduction of indigo and anthraquinone dyes to the soluble leuco-base form

Note that each of the dye molecules contains carbonyl groups (C=O). It is at these carbonyl sites that the reduction takes place. Dithionite (sodium hydrosulfite, Na$_2$S$_2$O$_4$) is generally used as the reducing agent and is itself oxidized in the reaction. (See Figure A5.)

$$Na_2S_2O_4 + H_2O \xrightarrow{[O]} 2NaHSO_3$$

sodium
hydrosulfite

sodium
bisulfite

Figure A5: Oxidation of sodium hydrosulfite

Dyeability of Various Fibers

In the case of vat dyes, the dyeability of a fiber primarily depends upon its ability to absorb the dye molecules into the strands of the fiber. This relates to the porosity of the fiber. Vat dyes are generally well absorbed by cellulosic fibers, such as cotton, because of the relatively open structure of the cellulose polymer.

Cellulose

Cellulose is a condensation polymer of beta-D-glucose (ß-D-glucose) units which combine by a dehydration process. This loss of water molecules, shown in the boxes in Figure A6, occurs between the hydroxyl groups (–OH) in adjacent ß-D-glucose molecules. Alternate ß-D-glucose molecules are inverted.

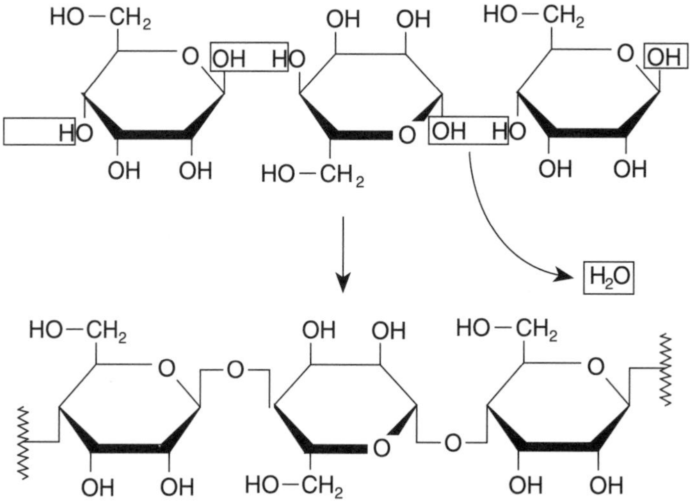

Figure A6: Condensation polymerization of β-D-glucose

There are about 10,000 glucose monomers forming each unbranched chain of cellulose. In the cellulose fiber, the cellulose chains exhibit both crystalline and amorphous regions. The cellulose chains are parallel to each other in the crystalline regions and more randomly spaced in the amorphous regions. (The dyeing takes place only in the amorphous regions.) Cross-linking is quite extensive because of the hydrogen bonding between the ß-D-glucose monomers on adjacent chains. (See Figure A7.) This three-dimensional attraction provides much mechanical strength to the network and accounts for the structural role that cellulose plays in plants, including cell walls and stems.

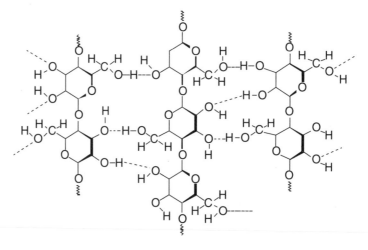

Figure A7: Hydrogen bonding (shown by dashed lines) exists between the cellulose chains.

Figure A8 provides a simplified diagram of a single cellulose layer with the trapped vat dye molecules (represented by the shaded ovals). The glucose monomers in the cellular chains are represented by hexagons and dotted lines show hydrogen bonds between the polymer chains.

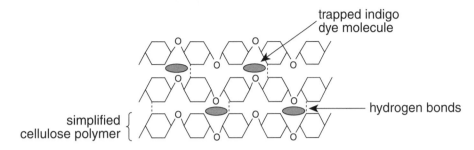

Figure A8: Vat dye molecules in a cellulose fiber

Synthetic Fibers

In contrast to cotton, synthetic fibers have a very tightly packed structure, which makes it difficult for even small molecules (including water) to penetrate the fiber under normal conditions. Polyethylene glycol terephthalate is the major commercial polyester textile fiber. It is a condensation polymer made by reacting ethylene glycol ($C_2H_6O_2$) with terephthalic acid ($C_8H_6O_4$). (See Figure A9.)

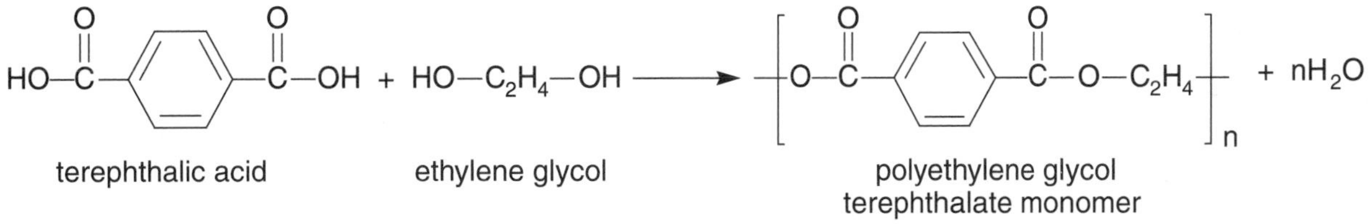

terephthalic acid   ethylene glycol   polyethylene glycol terephthalate monomer

Figure A9: Condensation polymerization reaction of polyethylene glycol terephthalate

In polyester, polymer chains are packed tightly side by side (See Figure A10) and, because of their close proximity, are held together by Van der Waals forces. Thus, polyester fibers have much less open space than cotton. The compactness of the polyester fibers compared to that of cotton is reflected in their densities. The density of polyester is 1.78 g/mL, compared to that of cotton at 1.05 g/mL.

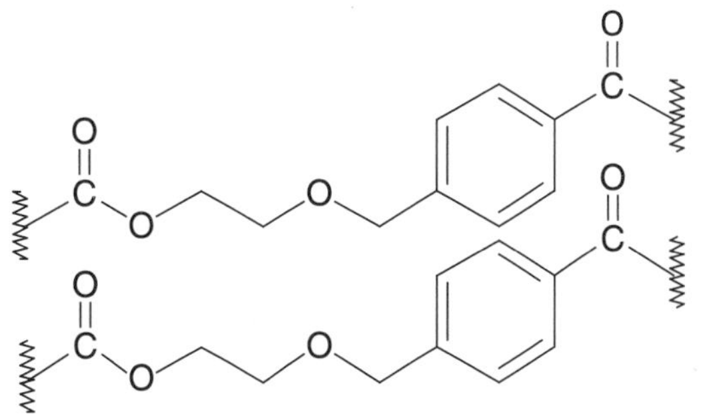

Figure A10: Alignment of polyester polymer chains

Special dyeing techniques are required to dye these synthetic fabrics. Typically these special techniques utilize small dye molecules and involve heating, which opens or loosens the fiber's structure to allow the dye to temporarily penetrate the fiber. After the dyeing, the fiber reverts back to its closely packed state, trapping the dye molecules in the fiber.

■ Notes and Setups for Activities

This section provides a materials list and safety and disposal information for each activity. Answers to Summary Questions are also included.

Activity 1: What Is Denim?

Each student will need a 3-cm x 3-cm square of denim fabric and a magnifying lens. Be sure to obtain the sized variety of denim for this activity. (It is the "stiff" denim.)

Activity 2: How Is Synthetic Indigo Prepared?

This activity is an organic synthesis done on a very small scale. On this scale, the amounts of reagents used are so small that the cost and disposal problems are minimized. In addition, the amount of time for the reaction to take place is minimized so that it is practical to carry out the synthesis during a 45-minute laboratory period.

Safety and Disposal

Dust, pellets, and solutions of sodium hydroxide (NaOH) are very caustic. NaOH can cause severe chemical burns and destroy cell membranes. Contact with the skin and the eyes must be prevented. Should contact occur, rinse the affected area with water for 15 minutes. If the contact involves the eyes, medical attention should be sought while rinsing is occurring. Preparation of the NaOH solution is very exothermic. Have an ice bath available. (Tap water may be used.) When preparing this solution, add the solid pellets to the distilled water. When all of the pellets are dissolved and the solution cools somewhat, it may be diluted to the desired final volume with distilled water. Eye protection is required at all times when making or handling this solution. The unused NaOH solution can be saved for future use or diluted with water and flushed down the drain.

Acetone is flammable. Keep flames away. Hot plates should be used for heating the hot-water baths.

Indigo stains hands and clothing. Students should wear aprons to protect clothing. Lab gloves may be worn to protect hands. Supernatant liquid from the preparation of indigo can be diluted with water and flushed down the drain.

Materials

Per class of 24 students working in pairs
- 5.0 g ortho-nitrobenzaldehyde
- 20 mL acetone
- 15 solid sodium hydroxide (NaOH) pellets
- tweezers or forceps to handle the NaOH pellets
- 6 hot plates

- tap water
- ice
- distilled water

Per group
- 2 250-mL beakers
- 2 small test tubes (13 mm x 100 mm)
- small spatula
- test tube holder
- 10-mL graduated cylinder
- small funnel
- filter paper
- thermometer
- small zipper-type plastic bag and label
- apron for each student
- (optional) lab gloves for each student
- goggles for each student

Answers to Summary Questions

1. Indigo was traditionally obtained from plants—the *Indigofera* plant in India and Asia and the woad plant in Europe.

2. Solid indigo is blue-black with a metallic sheen. The metallic sheen is a common characteristic of minerals so indigo might easily be mistaken as being of mineral origin.

3. A great deal of indigo is used commercially as a dye and sufficient amounts could not be obtained from natural plant sources to satisfy the demand.

Activity 3: How Is Indigo Used as a Vat Dye?

This activity is written as a laboratory activity but could be done as a demonstration by using larger quantities and dyeing a piece of cloth instead of a small strip of fabric.

Safety and Disposal

Dust, pellets, and solutions of sodium hydroxide (NaOH) are very caustic. NaOH can cause severe chemical burns and destroy cell membranes. Contact with the skin and the eyes must be prevented. Should contact occur, rinse the affected area with water for 15 minutes. If the contact involves the eyes, medical attention should be sought while rinsing is occurring. Preparation of the NaOH solution is very exothermic. Have an ice bath available. (Tap water may be used.) When preparing this solution, add the solid pellets to the distilled water. When all of the pellets are dissolved and the solution cools somewhat, it may be diluted to the desired final volume with distilled water. Eye protection is required at all times when making or handling this solution. The unused NaOH solution can be saved for future use or diluted with water and flushed down the drain.

Indigo stains hands and clothing. Students should wear aprons to protect clothing. Lab gloves may be worn to protect hands. Indigo solutions may be diluted with water and flushed down the drain for disposal.

Some people are sensitive to the odor of dithionite. Use it in a well-ventilated area.

Materials

Per class of 24 students working in pairs
- 20 mL approximately 7.5 M sodium hydroxide solution, NaOH(aq), made from
  - approximately 6 g solid NaOH pellets
  - distilled water
- 2.0 g commercial Vat Blue Indigo, solid
- 3.0 g dithionite (sodium hydrosulfite, or sodium dithionite, $Na_2S_2O_4$), solid
- 6 hot plates
- tap water

Per group
- 250-mL beaker
- 100-mL beaker
- small test tube (13 mm x 100 mm)
- test tube holder
- disposable plastic transfer pipet
- tweezers or forceps
- strip of prewashed cotton fabric approximately 3 cm x 10 cm
- thermometer
- small spatula
- 10-mL graduated cylinder
- distilled water
- glass stirring rod
- (optional) lab gloves for each student
- apron for each student
- goggles

Getting Ready

Prepare a 7.5 M sodium hydroxide solution (NaOH) by dissolving approximately 6 g solid NaOH pellets in 20 mL distilled water. Use caution in preparing and handling. (See Safety and Disposal.) When preparing this solution, add the solid pellets to water. When all of the pellets are dissolved and the solution cools somewhat, it may be diluted to the desired final volume with distilled water.

The beaker and NaOH solution may be hot! The flask or beaker must be placed in an ice bath.

Answers to Summary Questions

1. Indigo is water insoluble. In the leuco-base form, it becomes soluble in water and will dissolve in the dye bath.
2. In the dye bath, the fabric appears light yellow-green.

3. As the fabric is exposed to air and dries, it becomes blue.
4. Answers will vary—generally the blue color will be developed after 10–15 minutes.
5. The indigo in the leuco-base form was oxidized by the oxygen in the air.
6. Indigo is the "blue" in your blue jeans. The dithionite would reduce it to the colorless leuco-base form so the jeans would lose their typical blue color.

Activity 4: What Is the Best Way to Develop an Inkodye Color?

Inkodye is a soluble vat dye, which means that it is preprocessed into the leuco-base form. It is sometimes available in art supply or craft stores or may be ordered from: Screen Process Supplies Manufacturing Company, 530 MacDonald Ave., Richmond, CA 94801.

Safety and Disposal

Eye protection should be worn during this experiment. If a UV light is used, warn students not to look directly at the light source. All excess solutions can be flushed down the drain with water.

Materials

Per class of 24 students working in pairs
- 24-well plastic microscale reaction plate (A 50-mL beaker may be substituted.)
- 2 disposable plastic transfer pipets
- toothpicks
- strips of cotton fabric approximately 1 cm x 6 cm for each test
- Inkodye colors
- tap water
- index cards
- stapler
- goggles
- 250-mL beaker with soapy water
- 250-mL beaker with clear water
- access to light and heat sources such as:
  - direct sunlight
  - indirect sunlight (through window)
  - incandescent light
  - fluorescent light
  - UV (black) light
  - hair dryer
  - oven (275°F)
  - hot iron (cotton setting)

Answers to Summary Questions

1. Optimum conditions for developing Inkodye colors appear to be either bright sunlight or a hot oven. In both cases the color developed rapidly and fully. Other light or heat sources either did not develop the color fully or were much slower to develop the color.

2. The colorless leuco-base form of a vat dye changes to the colored form when it is oxidized. Oxygen in the air is the oxidizing agent in this reaction.

3. It appears that air is needed as the oxidizing agent. Light may promote the reaction but is not necessarily essential, as the color of Inkodye also developed when heated in a closed, dark oven.

4. Designs will vary but should include some method for application of Inkodye to the fabric and some heat or light exposure to develop the Inkodye color.

Activity 5: Is Inkodye Equally Effective on All Types of Fabric?

The fabric strips for these tests should be pure samples, not blends. Yard goods can be purchased at fabric stores or commercial multifiber test strips can be ordered from: Shirley Developments Ltd., Machine Control B.A.A. CANADA INC., 701 Ave. Meloche, Dorval, Quebec H9P 2S4.

Inkodye is a typical vat dye in terms of its reactions with various fabrics. The colors are developed by air oxidation which is speeded by exposure to sunlight or heat.

Materials

Per class of 24 students working in pairs
- 24-well plastic microscale reaction plate
- 2 disposable plastic transfer pipets
- fabric samples (1 cm x 6 cm) including:
  ○ cotton
  ○ linen
  ○ nylon
  ○ Dacron™
  ○ polyester
- 1 index card for each fabric sample
- toothpicks
- Inkodye
- access to direct sunlight or an oven
- stapler
- 250-mL beaker with soapy water
- 250-mL beaker with clear water
- goggles for each student

Answers to Summary Questions

1. Cotton and linen are both made up primarily of cellulose, 95% and 82% respectively, so one would expect them to react similarly.

2. Synthetic fibers have a very tightly packed structure with their polymer chains aligned side by side.

3. Synthetic fibers are not well dyed by vat dyes because of their tightly packed fiber structure; the relatively large dye molecules cannot penetrate the fibers. Cellulose is a linear polymer with a fairly open structure that allows molecules of the vat dye to enter the fiber between the polymer chains.

■ Supplementary Activities

Literature Integration

Have students read and report on one of the following:

- "Indigo," by Nicholas Carnac (St. Martin's Press, New York, NY, 1982, ISBN 0-312-41413-7). This is a novel set in the indigo plantations of India.

- "Blue Goes for Down—A Liberian Folk Tale," recorded by Ester Dendel. (Text provided here.) This is a Liberian legend about how the properties of indigo were discovered. Have students compare the water spirit's instructions to Asi with the procedure they used to make indigo and dye fabric. Discuss the importance of the color blue (and thus indigo) to the people of Foya Kamara and the people of 20th-century America.

Blue Goes for Down

How indigo dye came to Liberia—a folk tale recorded by Ester Warner Dendel

In the long ago and far away when High God left the earth, he went to live in the sky. The sky was close to earth in those days, so close it rested on the hills and mountains and sagged into the valleys. Energetic women feared to beat their pestles too high lest they pierce the fabric of the sky just above their heads and poke the spirit of a departed elder. What calamity!

It was better, really better, that High God, after being whacked a few times by busy women, left the spirits of the departed elders and went higher and farther from people. At least the low-lying sky was left to blanket man and shield him from the fierce sun. The people in their loneliness for God made sacrifices to the spirits of the ancestors and gave them messages to carry to God.

The sky did more for man in those days than to shade him and to house the spirits. Bits of sky could be eaten. This was different from other foods. Rice and palm oil fill the belly. Sky fills the heart. With a scrap of cloud inside him, a person can float and dream and find again the peaceful, joyous feelings that filled him before High God left the earth.

It was a dangerous business, this eating of cloud. One had to come to cloud-food pure in thought and body. Even so, one could become cloud-drunk, sweetly drunk and unknowing. This is what happened to Asi, the seeress of Foya Kamara.

On a bright morning Asi came to the banks of the stream that flows past the town. She came with her girl child tied on her back under a pure white lappa of country cloth. On Asi's head was a raffia bag filled with rice which she must cook and eat on the sacred spot where an altar to the river spirit stood against a great silk-cotton tree. In her hands she held a hollow stick. In its hollow was the winking red eye of a lump of charcoal for lighting the sacred fire.

Asi walked calmly, her head high and straight as she neared the altar because one does not rush with unseemly haste to a sacred place. She collected sticks from the forest and lighted the fire between three rocks which held the sacred clay pot which was always left in the forest. After she had spread her lappa on the earth and made a cushion of leaves under it to soften the place for her child, she walked without clothes to the bank of the stream where she would rinse the pot and take water for cooking.

On sunny days strips of cloud came to lie down in the river. One could look down into the deep pools and see the beautiful blue color of the sky lying there in the sacred wetness. Asi had eyes and heart that were hungry for color. To Asi, the blue of the pools was the most beautiful color in all the world. Asi looked back at the bank of the stream where her child was lying on the white lappa. The color of the white lappa seemed a dead and lifeless thing that had never known sun or cloud or sky.

"Perhaps," thought Asi, "if I eat enough sky, the blue will come to my skin from inside me. With luck, my hair will be thunder-blue."

Asi shivered then because she knew that a seeress must not beg anything for herself at the holy pools; one must ask only for the entire people of the village. She had done a selfish, wicked thing just when she should have been most pure in her heart. Fear shook her body as she carried water for the rice toward the fire. What was done was done, the wicked thought had taken hold of her, she must beg forgiveness of the water spirit and think now of her sacred task.

When the pot of water had been set above the fire, Asi sat down with her back against the great silk-cotton tree, waiting for the water to boil. "I will eat some sky now to make my heart lie down and be still," Asi told herself. Reaching up, she broke off a strip of sky as long as a plantain leaf and began to feed her lonely heart.

With the first swallow of sky, beautiful thoughts filled Asi. She felt herself within the roots of the trees far below her in the river-wet soil. The roots nuzzled the earth to drink the holy wetness the way a baby nuzzles a mother's breast to find milk. Asi's own breasts ached with the nuzzling of the roots because her spirit was there inside the sacred roots.

When the roots had drunk their fill and were ready to sleep, Asi's spirit rose and entered the body of a veda bird dancing in the air before her. The veda is a blue so bright it is a hurting, a lovely hurting to the eyes. It dances in one spot in the air when it is ready to mate. It was from floundering in the sky where the blue rubbed off on its body that the veda became this trembling, beautiful blue color. Perhaps if she asked for the blue for all the people, not just for herself…. Asi rose and added the rice to the water in the pot which had begun to boil. She was calmer now and not so afraid since she had decided to make a begging for blue to come down to all the people of Foya Kamara. She saw that her baby was asleep on the white lappa. Asi was free to eat just one more bit of sky while the rice cooked. She would then leave her begging for blue along with some rice on the altar and go home before the forest was dark.

When Asi awoke, her head throbbed and she knew she had been drunk with sky. The forest no longer smelled sweet. No birds sang. In her nostrils was the stench of burned rice; she had spoiled the sacrifice she had come to make. The sun was low in the sky. Fear ate at Asi when she turned her aching head to look to her child. The baby had rolled off the lappa and was lying face down on the earth. Something strange about the lappa caught Asi's eye; there was a blue patch of color in the center where the baby had wet. One small patch of deep blue in the dead expanse of white. Asi did not stop to finger the lappa. She rose to her feet as quickly as she could get her joints together and ran to her baby. When she turned the child over, no breath came from its mouth.

Asi's baby was dead. This was the punishment for bringing selfish thoughts to that holy place. In a frenzy of grief Asi ran to the fire, now dead ashes, and loosened her hair to receive the grime of the ashes as is the custom with women in mourning. Tears streamed down her face, streaking the ashes she had piled on her head. Asi clutched her child to her, then wrapped the lifeless body in the lappa which was her own skirt. Her body rocked forward and back as she wept.

Finally, Asi felt the life and the grief going out of her. She fainted there at the base of the silk-cotton tree. And while she was in faint, the water spirit spoke to her, telling her about the blue spot on the white lappa. It was indigo, the spirit told her, and came from the leaves she had plucked to cushion her child. In order for the blue to stay, there must be urine and salt and ashes with indigo. It was necessary for the baby's spirit to leave its body; otherwise, Asi would not have added the salt of her tears and the ashes of her grief; the blue Asi had desired above all else would not have stayed on the earth.

Before Asi awakened from this trance, the spirit cautioned her that now since the color blue had come down to earth to stay, it was a sacred duty to guard the indigo and that only women too old to bear children should handle the indigo pots. Asi was to carry her new knowledge back to Foya Kamara and instruct the old women there how to make the blue juice live happily in the cloth for all the people. Only after that would Asi conceive again and the spirit of her child, just dead, return to live in her hut.

When the people of Foya Kamara awoke the next morning, they saw that the sky no longer rested on the hills or sagged to the roofs of the houses. High God, after having let women have the secret of blue for their clothes, pulled the sky up higher where no one could reach up to break off a piece for food. People look on the blue of fine cloth and have less need of a near sky, even though in their hearts they will always remain lonely for God.

Source: Natural Plant Dyeing, *published by the Brooklyn Botanic Garden, Brooklyn, N.Y., 11225 (published at 2601 Sission Street, Baltimore, MD)*

Other Cross-Curricular Integration

The following activities are suggestions for integrating the scientific study of vat dyes with other disciplines.

History and Economics

Have students research and report on the history of the dye industry; the influence of indigo importation to the economics of the woad industry in Europe; and later, the influence of synthetic indigo on the indigo trade. Have students study and report on commercial trade routes and their influence on the spread of dye use from one continent to another. Have students research and report on the influence of the Revolutionary War on the indigo industry in colonial America.

Art

Have students conduct major art projects using Inkodye as the medium.

Painting Technique
Inkodye may be used to paint in a direct, spontaneous fashion resembling watercolor technique. One fascinating approach is to paint with the fabric spread in hot sunlight. The colors and the painting, as a whole, will develop as the work progresses.

Tie-Dye Patterns
Tie dyeing with just one color of Inkodye, followed by a sunlight or baking method of development, produces intricate and interesting color gradations. When dyeing with two colors, the following procedure is suggested:

1. Tie approximately ⅔ of the area.
2. Apply the lighter-colored dye, and develop.
3. Untie half the ties.
4. Tie the previously untied area.
5. Apply the darker-colored dye, and develop.

Special tie-dye effects are obtained by the following:

- Apply dye into crevices of fabric with a squeeze bottle.
- Spoon on a small amounts of dye to color local areas.
- During dipping, protect some areas from the dye with tied plastic bags.

Sundry Application Methods
Distinctive patterns can be produced by various methods of application, such as:

- Draw lines with a squeeze bottle.
- Throw splatters from a brush.
- Press mosaic patterns with a sponge.
- Make leaf patterns by painting the back of a leaf and pressing it onto the fabric.

■ References

Books

The Encyclopedia of Crafts; Torbert, L., Ed.; Scribner's: New York, 1980.

Gordon, P.F.; Gregory, P. *Organic Chemistry;* Springer-Verlag: New York, 1983.

Inkodye; Screen Process Supplies Manufacturing Company: Oakland, CA.

Linton, G.E. *Modern Textile and Apparel Dictionary;* Textile Bookservice: Plainfield, NJ, 1973.

Natural Plant Dyeing; Weigle, P., Ed.; Brooklyn Botanic Garden: Brooklyn, NY, 1973.

Needles, H.L. *Textile, Fibers and Dyes;* Noyes: Park Ridge, NJ, 1986.

Trotman, E.R. *Dyeing and Chemical Technology of Textile Fibers;* Griffin: England, 1984.

Journals

Fernelius, W.C.; Renfrew, E.E. "Indigo," *Journal of Chemical Education.* 1983, 60(8), 633–634.

McKee, J.R.; Zanger, M. "A Microscale Synthesis of Indigo: Vat Dyeing," *Journal of Chemical Education.* 1991, 68(10), A242–244.

Part B: Classroom Materials

What Makes Blue Jeans Blue?
> Activity 1: What Is Denim?
> Overhead 1: The Woven Structure of Denim
> Overhead 2: Ring-Dyed Threads in Denim

Historical Setting
> Student Background 1: A Brief History of Dyeing

What Are Vat Dyes?
> Student Background 2: The Chemistry of Vat Dyes
> Overhead 3: Oxidation-Reduction Reaction of Vat Dyes

What's a "Mood Indigo?"
> Student Background 3: Indigo
> Overhead 4: Indigo Structure
> Activity 2: How Is Synthetic Indigo Prepared?
> Activity 3: How Is Indigo Used as a Vat Dye?

What's New in Vat Dyeing?
> Activity 4: What Is the Best Way to Develop Inkodye Colors?

Dyes and Structures of Fibers
> Student Background 4: Textile Fibers and Dyes
> Overhead 5: Fiber Polymer Structures of Cellulose and Polyester
> Activity 5: Is Inkodye Equally Effective on All Types of Fabric?

What Makes Blue Jeans Blue?

Activity 1: What Is Denim?

The history of denim goes back about 1,600 years to the city of Nîmes, France, where it was first known as "serge de Nîmes." "De Nîmes" (literally, "of Nîmes") later became Americanized to "denim." This rugged cotton fabric, originally associated with typical work clothes of French peasants, appeared in the U.S. sometime between the War of 1812 and the Civil War as covering for the famed Conestoga wagons that carried pioneers on the long trek to the West.

The usual blue color of denim results from the use of indigo as a dye. Indigo belongs to the family of dyes known as vat dyes. The insoluble indigo must first be reduced to a soluble form so that it can dissolve in the dye bath and be used to wet the fabric. The indigo particles in their soluble form are attracted to the cellulose fibers in the denim, most likely because the dye molecules are the correct shape to fit between the chains of cellulose that make up the fiber. Upon exposure to air, the soluble form of the indigo is oxidized to the insoluble form. In the process the large, insoluble indigo molecules become trapped between the cellulose chains. Generally, the dye penetrates only the outer layer of the thread, resulting in a ring-dyed thread. (See Figure B1.)

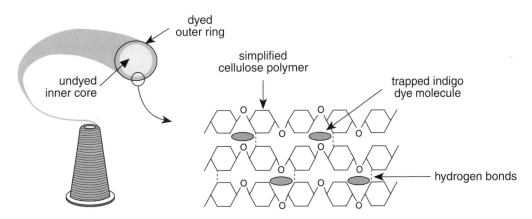

Figure B1: Ring-dyed thread showing indigo trapped in cellulose polymer

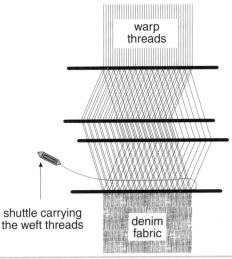

When denim is woven on a loom, the vertical threads, called the warp, are pulled taut. Because of the stress associated with this process, the warp threads are precoated with starch to provide additional strength. Textile manufacturers call this pretreatment "sizing." The threads that are woven horizontally, over and under the warp threads, are known as the weft. The weft threads are typically not sized. A shuttle is used to carry the weft back and forth through the warp. (See Figure B2.)

Figure B2: Denim fabric being woven on a loom

Examination of a Denim Sample

Using a magnifying lens, carry out the following examinations on a small square of denim fabric.

1. Look closely at the denim with the magnifying lens. Examine both sides of the fabric. How are they different? With the darker side of the denim facing you, place the square so that the diagonal pattern of the weave goes from bottom left to top right, as in Figure B3.

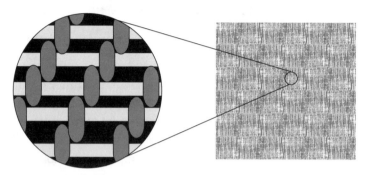

Figure B3: Denim sample with close-up of weave

2. Pull one thread away from the right-hand edge of the fabric. (See Figure B4a.) What color is it? Remove five more threads from the right-hand edge.

3. Pull one thread away from the top edge of the fabric. (See Figure B4a.) What color is it? Remove five more threads from the top edge.

4. Examine the row of blue threads at the top of the sample with the magnifying lens, looking at the ends of the threads. Has the dye completely penetrated each thread or is a white core apparent in the center of the thread? (See Figure B4b.)

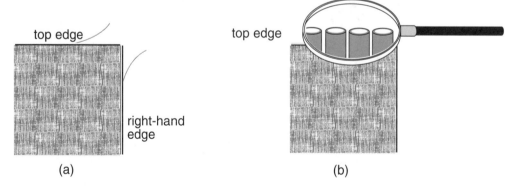

Figure B4: (a) Remove threads from the denim sample. (b) Use a magnifying lens to examine the row of blue threads at the top edge of the sample.

Figure B5: Droop test with denim threads

5. Select one of the blue threads and one of the white threads you have removed from the denim and conduct a "droop test". Hold the threads vertically between your thumb and forefinger. (See Figure B5.) Which one "droops?" Which threads in the denim are the warp, and which are the weft? By carefully removing threads from the fabric, determine the weaving pattern. How many white threads does each blue thread cross and vice versa? Can you explain why the two sides of the denim square appear different?

What Makes Blue Jeans Blue?

Overhead 1: The Woven Structure of Denim

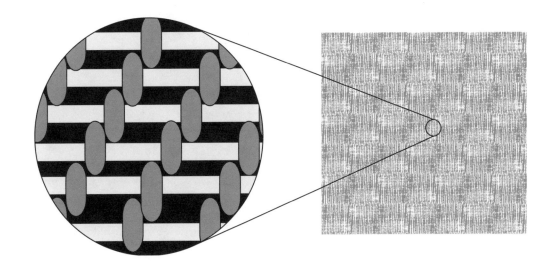

What Makes Blue Jeans Blue?

Overhead 2: Ring-Dyed Threads in Denim

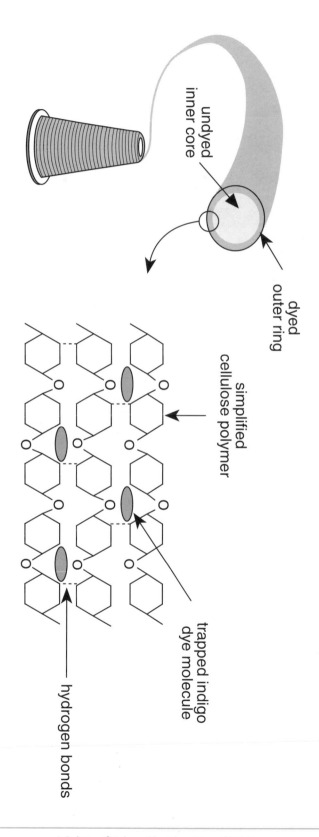

undyed
inner core

dyed
outer ring

simplified
cellulose polymer

trapped indigo
dye molecule

hydrogen bonds

A Palette of Color—The Chemistry of Vat Dyes

Historical Setting

Student Background 1: A Brief History of Dyeing

The ancient art of dyeing, practiced in Persia, China, and India thousands of years before the birth of Christ, was most likely brought to Egypt by Phoenician traders. Egyptian tombs at Thebes have revealed the earliest dyed textiles, dating back to 3500 B.C., which show traces of indigo dye. Indigo was an important dye in the ancient world. Romans had it imported at great expense from India and named it Indicum (after its source). Its name evolved to indican and finally to indigo.

Indigo is derived from the juice of the plant *Indigoferae sumatrana.* For many years the imported crystalline indigo cakes from India were assumed to be of mineral origin. The use of indigo did not spread into northern and western Europe until the 16th and 17th centuries. In Europe, indigo was extracted from the woad plant, *Isatis tinctoria.* The indigo plant is much richer in dye content than woad and the imported material began to threaten local production. In order to protect the woad industry, the English government imposed a severe tax on imported indigo during the 1700s.

The Phoenicians ruled the dye industry from the 15th century B.C. until 638 A.D., when conquering armies destroyed the dye works. The Phoenician and Alexandrine merchants were instrumental in bringing many dyestuffs to Greece. From there the dyestuffs, their cultivation, and dyeing methods slowly found their way to Western Europe via Italy.

In 1540 Giovanni Ventgur Rosetti of Venice published *Plictho del Arti de Tentori,* a book containing 217 dye recipes and techniques for dyeing cloth and leather. It remained the primary source of information for dye makers for 200 years. Between the 16th and 18th centuries, dye makers in France and the Netherlands were considered the best in their trade. Many books and papers were written on the subject; however few secrets were revealed, as the books often contained rather vague instructions.

While efforts to improve dyes and dyeing were ongoing, it was not until 1856 that the first synthetic dye was developed by an Englishman, Sir William Henry Perkin. Perkin accidentally obtained the striking purple crystals of mauve while attempting to synthesize the anti-malarial drug quinine from coal tar. After establishing the usefulness of mauve as a dye, Perkin began manufacturing and selling it commercially. His success attracted many competent chemists to the dyestuffs industry and within 50 years of the initial discovery of mauve, over 90% of the dyes used were of synthetic origin.

Synthetic dyes replaced those from natural sources primarily because they were pure compounds which insured reproducibility of color. Today hundreds of synthetic dyes are available for use in a wide range of applications.

What Are Vat Dyes?

Student Background 2: The Chemistry of Vat Dyes

Oxidation and reduction reactions lie at the heart of vat dye chemistry. In the dyeing process as in many other oxidation-reduction systems, oxidation results from the combining of a molecule with oxygen, while reduction involves the removal of oxygen from a molecule. The two processes occur simultaneously.

Vat dyes are large, water-insoluble molecules. Treatment with a reducing agent (something which will remove oxygen from a compound) converts the dye molecule into a water-soluble form. Dithionite (sodium hydrosulfite, $Na_2S_2O_4$) in a basic solution is the most commonly used reducing agent for vat dyes. The water-soluble form of the dye is known as its leuco base. In this state, the dye molecule can be applied to a fiber in a water-dye bath. In the leuco-base form, vat dyes have an affinity for cellulose fibers, such as cotton. The dye molecules enter into the open spaces in the cellulose polymer. Drying the dyed fibers in air oxidizes the dye molecule back to the insoluble form. The insoluble dye particles are trapped inside the cellulose fibers, giving a permanent color to the fibers. (See Figure B6.)

$$\text{vat dye (insoluble)} \xrightarrow{\text{reduction}} \text{leuco-base form (soluble)} \xrightarrow{\text{oxidation}} \text{vat dye trapped in fiber (insoluble)}$$

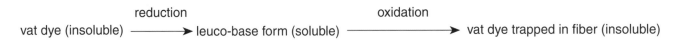

Figure B6: General process of solubilizing a vat dye

The term "vat dye" originated with the ancient processing of indigo. In order to extract the indigo from the leaves of the *Indigofera* plant, the leaves were fermented in large vats for several days. This "vatting" process was the reduction step in the chemical reaction. While vat dyes originally came only from natural materials, today most vat dyes are synthetically produced.

Most synthetic vat dyes belong to two chemical families, those related to indigo and those related to anthraquinone. (See Figure B7.) Note that both indigo and anthraquinone contain the characteristic carbonyl group (C=O), which is the site of the reduction and oxidation processes.

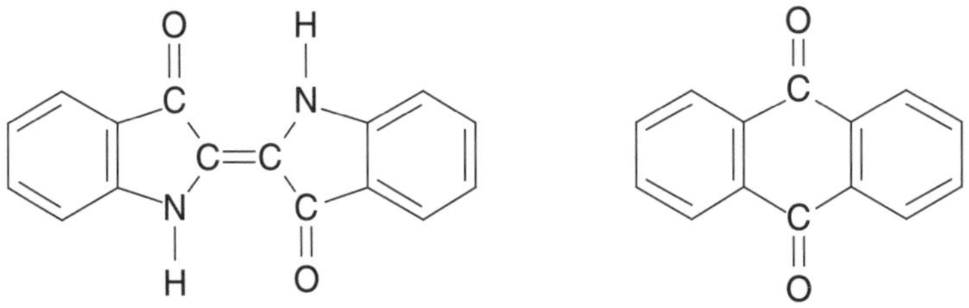

indigo anthraquinone

Figure B7: The structures of indigo and anthraquinone

■ What Are Vat Dyes?

Overhead 3: Oxidation-Reduction Reaction of Vat Dyes

$$\text{C=O} \xrightarrow[\text{NaOH}]{\text{Na}_2\text{S}_2\text{O}_4} \text{C}-\text{O}^{\ominus}\text{Na}^{\oplus} \xrightarrow{\frac{1}{2}[\text{O}_2]} \text{C=O}$$

water-insoluble
vat dye

soluble
leuco-base
form

regenerated
insoluble dye
trapped in fiber

What's a "Mood Indigo?"

Student Background 3: Indigo

If any one fabric could be used to characterize "style" in the United States, it would be denim. A recent industry survey estimated world demand to be 13,000 metric tons of blue denim per year. The dye behind this fabric is indigo. Perhaps the oldest dye known, indigo's preparation is described in ancient Sanskrit writings, and 5,000-year-old Egyptian mummies have been discovered wrapped in indigo-dyed cloth. Powdered indigo was also used as a cosmetic by the Greeks and Romans, who developed a blue indigo crayon to be used as eye shadow.

In Europe, leaves of the woad plant *(Isatis tinctoria)* were the source of indigo dye but the woad industry gradually died out as indigo from India and Southwest Asia was introduced around A.D. 1500. In India, the indigo was produced by fermenting the leaves of the *Indigofera* plant, which is much richer in dye content than woad.

Indigo was also produced in colonial America, with a million pounds exported from Charleston, South Carolina, in the late 18th century. However, this fledgling industry was ruined by the outbreak of the Revolutionary War.

Large indigo plantations in India provided the bulk of the dye during the 19th century. To meet the present-day demand for indigo, if it were still being produced by the original methods, half of India would have to be planted in indigo plantations. However, in the 1880s, Adolph von Baeyer (of aspirin fame), professor of chemistry at the University of Munich in Germany, undertook the work of preparing synthetic indigo. His success was rewarded with a Nobel Prize in Chemistry in 1905. The rights to his invention were purchased by Badische Anilin und Soda Fabrik (BASF) for $100,000. However, because of the difficulties in the manufacturing process, it was not until the 1930s that synthetic indigo essentially replaced natural indigo in the dye industry. By that time BASF had spent over $5 million to develop a commercially feasible process.

Some natural indigo is still produced today, especially on the African continent, where over 50 different plants have been found to contain indigo.

What's a "Mood Indigo?"

Overhead 4: Indigo Structure

What's a "Mood Indigo?"

Activity 2: How Is Synthetic Indigo Prepared?

The original title for jazz great Duke Ellington's "Mood Indigo" was "Dreamy Blue." Indigo is often used as a synonym for blue. Comedians have enjoyed the "pun potential" of Ellington's later title with one-liners such as: "There once was a cow that swallowed a bottle of blue ink and 'Mooed Indigo.'"

Indigo dye originally was a natural dye extracted from the leaves of the *Indigofera* plant; however, since the mid-1930s, most of the indigo used has been synthetically produced. In this experiment you will carry out an organic synthesis of indigo. The general reaction in this preparation is shown in Figure B8. The intermediate products in the reaction are not all known.

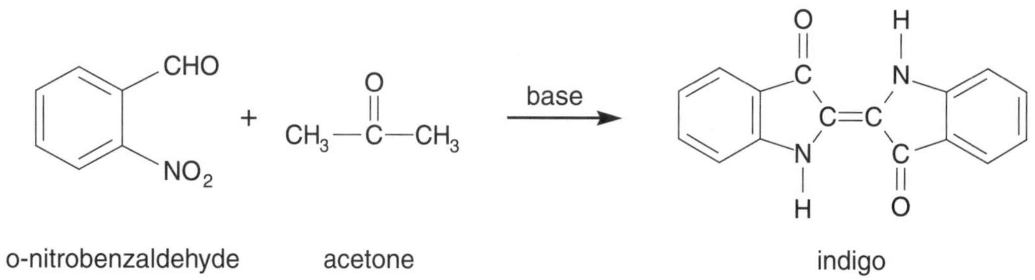

o-nitrobenzaldehyde   acetone   indigo

Figure B8: The general preparation reaction of indigo

Safety

Goggles are mandatory: wear them at all times during this experiment. Acetone is a flammable liquid, keep it away from open flames. To avoid open flames, a hot plate should be used to heat the water bath.

Sodium hydroxide (NaOH) pellets are caustic, handle them only with tweezers. In preparing the NaOH solution in Step 3, the test tube may become quite warm—care should be taken when handling. NaOH can cause severe chemical burns and destroy cell membranes. Contact with the skin and the eyes must be prevented. Should contact occur, rinse the affected area with water for 15 minutes. If the contact involves the eyes, medical attention should be sought while the rinsing is occurring.

Indigo will stain your hands and/or clothing. Wear an apron to protect your clothing. Lab gloves may be worn to protect your hands.

Materials

Per group of 2 students
- 2 250-mL beakers
- 2 small test tubes (13 mm x 100 mm)
- small spatula
- small funnel

- filter paper
- test tube holder
- 10-mL graduated cylinder
- thermometer
- zipper-type plastic bag and label
- hot plate (two groups may share the hot plate)
- ice
- tap water
- ortho-nitrobenzaldehyde
- distilled water
- solid sodium hydroxide (NaOH) pellets
- acetone
- goggles for each student
- apron for each student
- (optional) lab gloves for each student

Procedure

1. Prepare a hot-water bath by filling a 250-mL beaker ¾-full with tap water and warming it on a hot plate to about 50°C.

2. Using the graduated cylinder, measure 1 mL acetone into a small test tube and add ½ spatula of ortho-nitrobenzaldehyde. Shake the test tube gently to mix the chemicals and place the test tube into the hot-water bath.

3. Use tweezers to transfer 1 pellet of solid sodium hydroxide (NaOH) into a second small test tube. Add 1 mL distilled water and gently agitate the test tube to dissolve the NaOH.

The test tube may become warm.

4. Using the test tube holder, carefully remove the first test tube from the hot-water bath. Slowly, with periodic gentle shaking to mix the contents, add the aqueous NaOH solution from Step 3 to the acetone/ortho-nitrobenzaldehyde solution.

5. Return the test tube with the reaction mixture to the hot-water bath and heat for about 5 minutes, using the test tube holder to agitate the test tube occasionally.

6. Prepare an ice-water bath in a second 250-mL beaker by filling the beaker half full of a crushed-ice-and-tap water mixture.

7. Using the test tube holder, remove the test tube from the hot-water bath and add 2 mL distilled water.

The test tube will be hot and the solution containing hot NaOH is very caustic—handle with care.

8. Place the test tube with the reaction mixture in the ice-water bath and chill for 10 minutes.

9. Filter the reaction mixture using a filter paper cone in a small funnel, discarding the supernatant liquid.

10. Rinse the solid on the filter paper with small portions of distilled water, discarding the washings, and allow to air dry.

11. Place filter paper with the dry product (indigo) into a small zipper-type plastic bag and label.

The indigo product is a blue-black solid with a metallic sheen.

Summary Questions

1. In this laboratory activity you prepared synthetic indigo. What is another source of this dye?

2. Solid indigo was thought, at one time, to be a mineral. In fact, a mining permit for indigo was issued in England in the 17th century. Based on your knowledge of metal characteristics and the appearance of your product, explain how this confusion may have arisen.

3. Why is most of the indigo used today prepared synthetically rather than from natural plant sources?

What's a "Mood Indigo?"

Activity 3: How Is Indigo Used as a Vat Dye?

Indigo is an insoluble blue vat dye. In this experiment you will reduce the indigo to its soluble, colorless form (known as its leuco-base form) and use that solution to dye a fabric sample. The characteristic indigo color develops as the reduced, colorless form of the indigo oxidizes in air back to the insoluble blue colored form. (See Figure B9.)

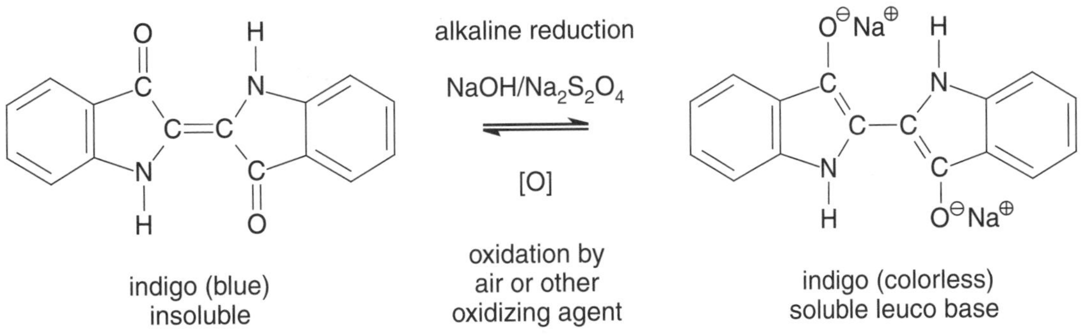

Figure B9: Reversible reduction-oxidation of indigo to its leuco base

Safety

Goggles are mandatory: wear them at all times during this experiment. Dust, pellets, and solutions of sodium hydroxide (NaOH) are very caustic. NaOH can cause severe chemical burns and destroy cell membranes. Contact with the skin and the eyes must be prevented. Should contact occur, rinse the affected area with water for 15 minutes. If the contact involves the eyes, medical attention should be sought while the rinsing is occurring.

Indigo will stain your hands and/or clothing. Wear an apron to protect your clothing. Lab gloves may be worn to protect your hands.

Materials

Per group of 2 students
- disposable plastic transfer pipet
- small spatula
- small test tube (13 mm x 100 mm)
- 250-mL beaker
- glass stirring rod
- thermometer
- 100-mL beaker
- tweezers or forceps
- test tube holder
- small graduated cylinder
- hot plate (two groups can share the hot plate)
- tap water
- strip of white cotton fabric approximately 3 cm x 10 cm

- distilled water
- goggles for each student
- 7.5 M sodium hydroxide solution, NaOH(aq)
- commercial Vat Blue Indigo, solid
- 3.0 g dithionite (sodium hydrosulfite, or sodium dithionite, $Na_2S_2O_4$), solid

Procedure

1. Prepare a hot-water bath by filling a 250-mL beaker ¾-full with tap water and warming it on a hot plate to about 50°C.

2. From the tip of a small spatula, place several grains of solid vat blue indigo into a small test tube.

3. Using a plastic transfer pipet, add 10 drops 7.5 M sodium hydroxide solution, NaOH(aq), to the test tube containing the indigo. Gradually, and with gentle periodic shaking, add 2 mL distilled water from the graduated cylinder to the mixture.

4. Warm the test tube containing the reaction mixture in a 50°C hot-water bath for 5 minutes. Grasping the warm test tube with the test tube holder, slowly add 1 small spatula full of dithionite ($Na_2S_2O_4$) to the test tube and gently shake to mix.

 Solutions of hot NaOH are very caustic. Handle with care.

5. Stirring frequently with a glass stirring rod, continue to warm the reaction mixture in the hot-water bath for approximately 5 minutes or until the reaction mixture turns a light yellow-green. This color indicates the formation of the leuco-base form of indigo.

6. Use a separate 100-mL beaker to heat 30 mL distilled water to 50°C. Stir several small crystals of dithionite ($Na_2S_2O_4$) into this 30 mL hot water. Remove the dilute $Na_2S_2O_4$ (aq) from the heat and allow to cool for several minutes.

7. Add the light yellow-green contents of the reaction test tube to the cooled dilute $Na_2S_2O_4$ (aq) in the beaker and stir briefly with a glass stirring rod.

8. Using tweezers, place a strip of prewashed cotton fabric into the dye bath. Allow the strip to soak in the dye bath for 6–8 minutes.

9. Remove the fabric from the dye bath with the tweezers, place it on a folded paper towel, and allow it to dry in the air.

Summary Questions

1. Why is it important to convert the indigo to the leuco-base form in the dye bath?

2. What color is the fabric while it is in the dye bath?

3. What color is the fabric once it dries?

4. How rapidly did the color change take place?

5. What is oxidized in this reaction? What is the oxidizing agent?

6. What would happen if you spilled dithionite onto your blue jeans?

What's New in Vat Dyeing?

Activity 4: What Is the Best Way to Develop an Inkodye Color?

Inkodye is the brand name of a group of vat dyes which have been preprocessed into the leuco-base form. In the leuco-base form, the dyes are almost colorless. Inkodye literature offers contradictory statements as to the best conditions under which to develop the Inkodye colors; both heat and light are cited as means of developing the dye. This activity examines various heat and light sources in order to determine the optimum conditions for developing Inkodye colors.

Safety

Eye protection should be worn during this experiment. Care should be taken with all heat sources. If a UV (black) light is used, avoid looking directly at the source.

Materials

Per group of 2 students
- 24-well plastic microscale reaction plate
- 2 disposable plastic transfer pipets
- toothpicks
- 250-mL beaker with soapy water
- 250-mL beaker with clear water
- goggles for each student
- index cards
- stapler
- strip of white cotton fabric approximately 1 cm x 6 cm for each test
- Inkodye
- heat and light sources such as:
    - direct sunlight
    - indirect sunlight (through window)
    - incandescent light
    - fluorescent light
    - UV (black) light
    - hair dryer
    - oven (275°F)
    - hot iron (cotton setting)

Procedure

1. In one well of a 24-well microscale reaction plate, mix 40 drops of water with 20 drops of the Inkodye color to be tested. Stir thoroughly with a toothpick.

2. Holding the 1-cm x 6-cm white cotton fabric sample by one end, dip it into the dye several times until the fabric is completely wet except for the end you are holding.

3. Lift the sample out of dye and allow the excess dye to drip back into the well.

4. Lay the sample on an index card and repeat the process with other samples, one for each light or heat source. Each sample should be placed on a different card.

5. Develop the color on the fabric samples by exposing each to different heat or light conditions. The following heat and light sources are suggested:

Light Sources	**Heat Sources**
Direct sunlight	Hair dryer
Indirect sunlight (through a window)	Oven (275°F)
Incandescent light	Hot iron (use cotton setting)
Fluorescent light	
UV (black) light	

Expose the samples to various sources for equal lengths of time. Label each index card with the heat or light source used.

6. After each sample has been developed, rinse it in soapy water and then in clear water to remove any excess reagent.

7. Staple each sample to its labeled card and complete the following data table.

Light or Heat Source	**Intensity of Color (light or dark)**
_____	_____
_____	_____
_____	_____
_____	_____
_____	_____

Summary Questions

1. Based on your experimental results, what are the optimum conditions for developing Inkodye colors?

2. How does the colorless leuco-base form of a vat dye change to the colored form? What role does air play in the process?

3. From your observations in this experiment, discuss the following statement which was found in a book on dyes and dyeing: "One particular brand of vat dyes (Inkodye) has been treated so that they are sensitive to light rather than to air."

4. As a textile engineer, it is your job to design a machine which will allow a piece of fabric to be dyed using Inkodye. Draw and discuss your design for such a machine, showing how the dye will be applied and how the color will be developed.

Dyes and Structures of Fibers

Student Background 4: Textile Fibers and Dyes

Textile fibers may be broadly classified into two types, natural and synthetic. Both types are composed of polymers (macromolecules made up of repeating monomer units).

Cellulose is the most abundant natural organic polymer. It occurs in all plants as a skeletal structure. Cotton is almost pure cellulose, containing up to 95%, while other vegetable fibers such as linen and flax contain a lower proportion. Linen, for example, contains 80–82% cellulose. Cellulose is a linear polymer of glucose. (See Figure B10.) The repeating monomer unit is two glucose units joined together. (See Figure B11.)

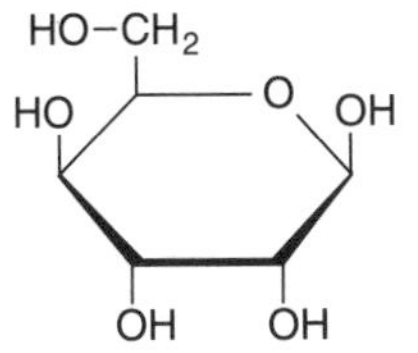

Figure B10: Glucose

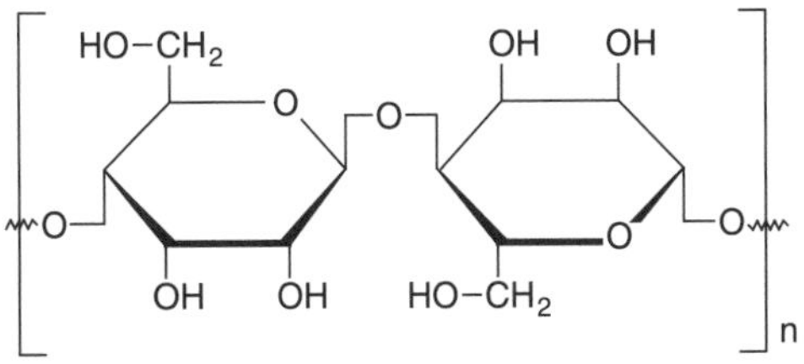

Figure B11: The monomer unit of cellulose

Approximately 1,200–1,300 glucose units are joined together to make the cellulose polymer chain. These long chains of glucose molecules are aligned parallel to one another. The chains are held together in a fairly open structure by extensive hydrogen bonding which occurs between the hydroxyl groups (–OH). (See Figure B12.)

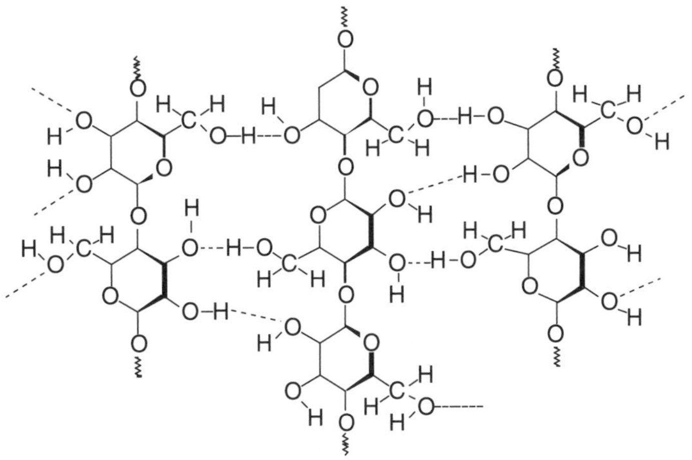

Figure B12: Hydrogen bonding (shown by dashed lines) in cellulose

A simplified diagram of the cellulose structure is shown in Figure B13 where the basic form of the glucose molecule is represented by a hexagon and the dotted lines represent hydrogen bonds between the polymer chains. In water, the cellulose polymer swells laterally, that is, the polymer chains move apart so that the structure becomes even more open, thus allowing the entry of vat dye molecules, shown by shaded ovals, between the polymer chains.

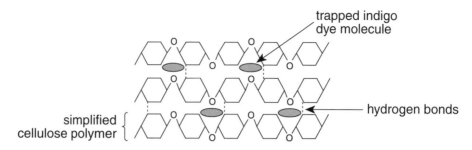

Figure B13: Vat dye molecules in a cellulose fiber

In contrast to cotton, synthetic fibers such as polyester have a tightly packed structure with the polymer chains aligned side by side. (See Figure B14.) These fibers have few open spaces, and as a result, even the smaller dye molecules have difficulty penetrating the fiber.

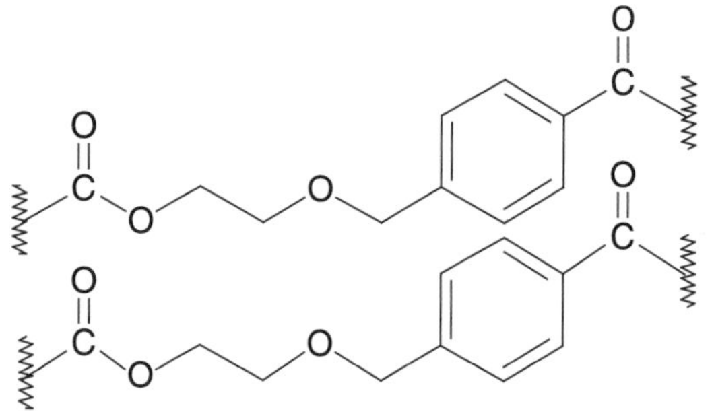

Figure B14: Alignment of polyester polymer chains

Dyes and Structures of Fibers

Overhead 5: Fiber Polymer Structures of Cellulose and Polyester

Polyester polymer

Simplified cellulose polymer

▪ Dyes and Structures of Fibers

Activity 5: Is Inkodye Equally Effective on All Types of Fabric?

The dyeability of a fiber by a vat dye such as Inkodye depends upon the fiber's ability to absorb the dye. Since vat dyes are not chemically bound to the fibers but are trapped between the polymer chains which make up the fiber, dye molecules must be able to enter the spaces between the polymer chains. The ability of a fiber to absorb the dye therefore is related to the amount of space between the polymer chains.

In this experiment you will test the ability of several fabric samples to absorb Inkodye.

Safety

Eye protection should be worn during this experiment.

Materials

For each group of 2 students
- 24-well plastic microscale reaction plate
- 2 disposable plastic transfer pipets
- 250-mL beaker with soapy water
- 250-mL beaker with clear water
- goggles for each student
- fabric samples (1-cm x 6-cm) including nylon, cotton, Dacron™, linen, polyester
- 1 index card for each fabric sample
- toothpicks
- stapler
- access to direct sunlight or an oven

Procedure

1. In one well of a 24-well microscale reaction plate, mix 40 drops of water with 20 drops of the Inkodye color to be tested. Stir thoroughly with a toothpick.

2. Holding a 1-cm x 6-cm sample of the fabric to be tested by one end, dip it into the dye several times until the fabric is completely wet except for the end you are holding.

3. Lift the sample out of the dye and allow the excess dye to drip back into the well.

4. Lay the sample on one half of a folded paper towel and label it appropriately.

5. Repeat Steps 2–4 with all the fabric samples you are testing. Be sure to identify each sample.

6. Develop the color of the Inkodye by placing the samples in direct sunlight for 10 minutes.

 Alternatively, the samples may be placed in a 275°F oven for 5–8 minutes.

7. Rinse the samples in soapy water and then in clear water to remove any excess chemical.

8. Staple the samples onto appropriately labeled index cards and complete the following data chart.

Type of Fabric	Absorption (good or poor)
_____	_____
_____	_____
_____	_____
_____	_____
_____	_____

Summary Questions

1. Study the structure of the fibers which make up your fabric sample. (See Student Background Handout 4.) Why did cotton and linen react similarly to the dye?

2. How do synthetic fabrics differ in structure from the natural fibers you tested?

3. How is the fiber structure of a fabric related to its ability to be dyed by Inkodye?